What is a Trilobite?

A Coloring Book by
The Georgia Mineral Society, Inc.

Written by Lori Carter

This edition published by:

The Georgia Mineral Society, Inc.
4138 Steve Reynolds Boulevard
Norcross, GA 30093-3059
www.gamineral.org

ISBN: 978-1-937617-10-3

What is a trilobite?

The Georgia Mineral Society, Inc.

Trilobites were creatures that lived a long time ago

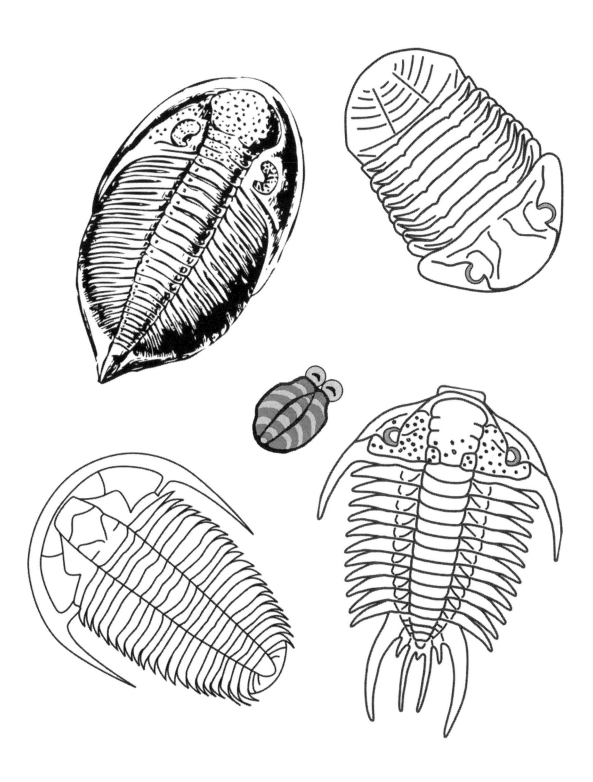

Do you think trilobites lived in trees?

Do you think trilobites lived in houses?

The Georgia Mineral Society, Inc.

Do you think trilobites lived in igloos?

Do you think trilobites lived in clouds?

Do you think trilobites lived in space?

Do you think trilobites lived in oceans?

Yes! Trilobites lived in oceans, but a long, long time ago

There were trilobites living on Earth for over 270 million years!

(From the Cambrian, 521 million years ago, to the Permian, 250 million years ago)

This is how to say trilobite

(try-low-byte)

The Georgia Mineral Society, Inc.

tri~~angle~~

+

~~s~~low

+

bite

The name "trilobite" means three lobes

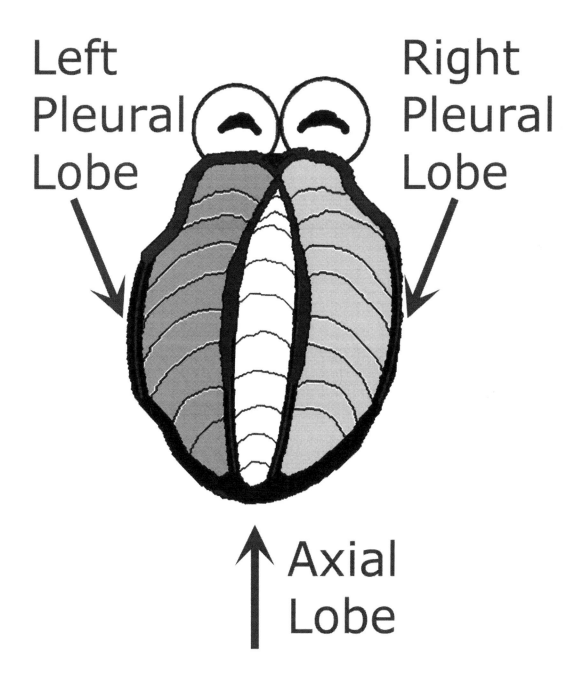

Left Pleural Lobe

Right Pleural Lobe

Axial Lobe

Trilobites have three distinct body parts too

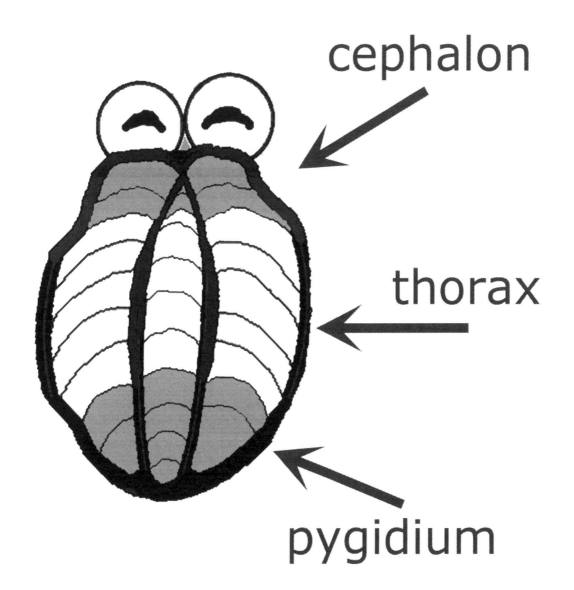

cephalon

thorax

pygidium

Trilobites were a type of creature called an "arthropod"

Arthropod skeletons are on the outside of their bodies

(These are not arthropods; they are just a silly way to depict exoskeletons)

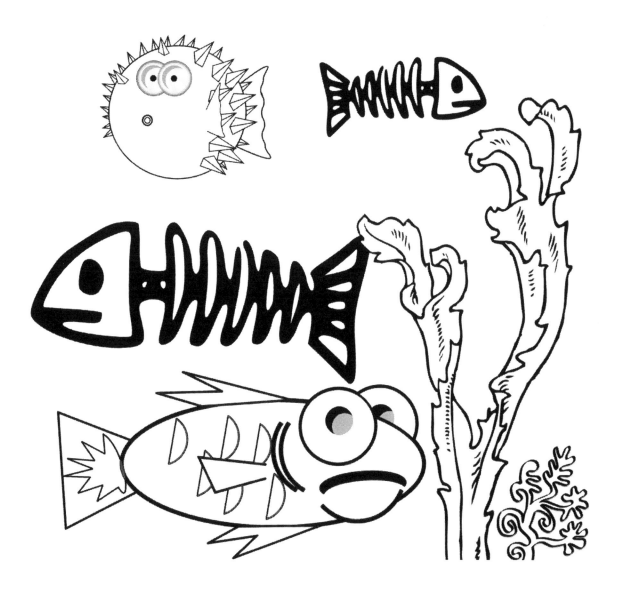

Arthropods have segmented bodies

segment

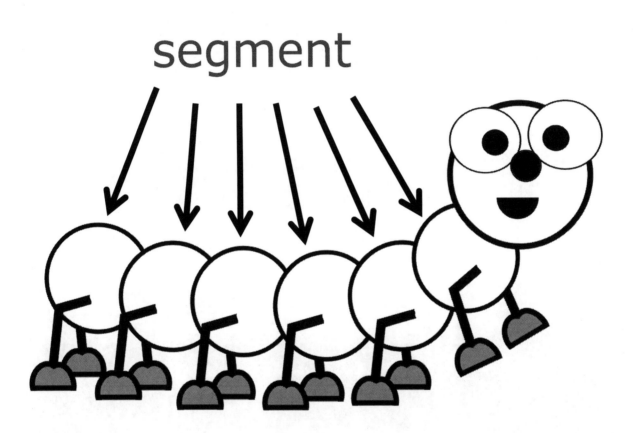

Arthropods have jointed appendages (arms, legs, etc.)

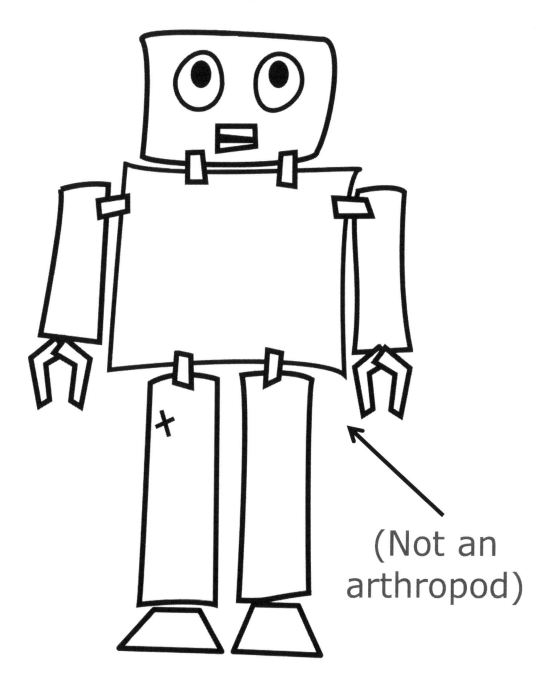

(Not an arthropod)

Insects are arthropods

Crustaceans are arthropods
(crabs, lobsters, shrimp...)

Arachnids are arthropods
(spiders, scorpions, ticks...)

Trilobites were many different sizes

The smallest trilobites were smaller than a flea

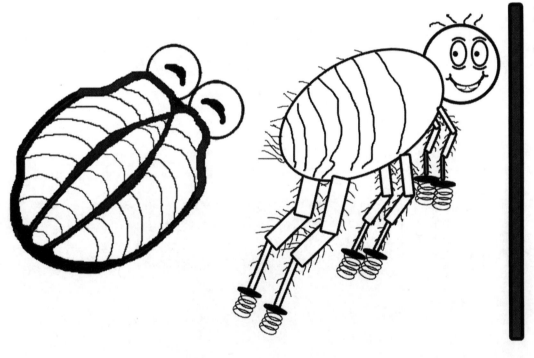

1 mm
.04 inches

The largest trilobites were about as long as a penguin is tall

70 cm
28 inches

Let's play some trilobite games!

Where did trilobites live?

In space
or

In igloos
or

in oceans?

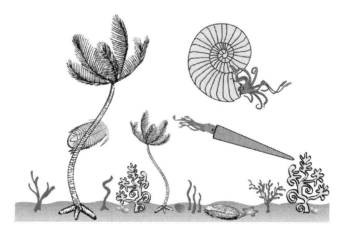

Find all of the arthropods !

Did you find all of the arthropods?

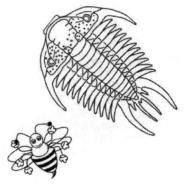

Fun Facts About Trilobites!

Some trilobites could roll into a ball to protect themselves similar to the way roly-polies roll up.

Electrolux, a company known for its vacuum cleaners, made one of the first commercially available robot vacuum cleaner. They called it "The Trilobite" and it was styled to look like a trilobite.

Some trilobites had eyes on the end of stalks. Most trilobites had compound eyes but some didn't have eyes at all.

The lenses of some trilobite eyes were made of calcium carbonate ($CaCO_3$), the main component of the shells of marine animals and eggshells.

Many of the trilobite fossils we find are actually molted exoskeletons.

Trilobite fossils have been found on every continent and sub-continent!

The Georgia Mineral Society, Inc.

Made in the USA
San Bernardino, CA
23 December 2017